D1239152

Curious Kids Guides

INSECTS AND BUGS

Amanda O'Neill

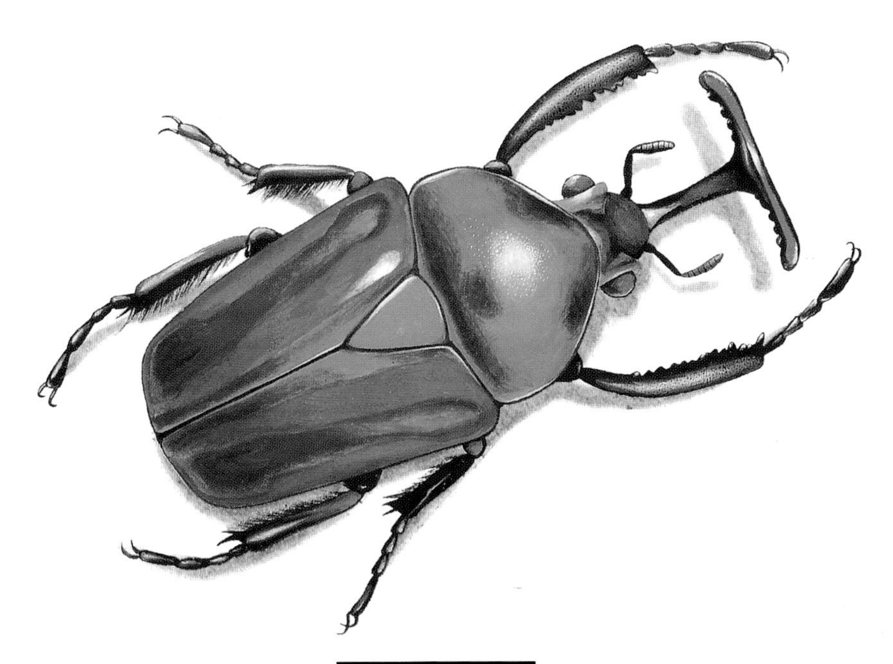

KINGfISHER

NEW YORK

INSECTS

KINGFISHER
Larousse Kingfisher Chambers Inc.
80 Maiden Lane
New York, New York 10038
www.kingfisherpub.com

First published in 1994
First published in this format 2002
10 9 8 7 6 5 4 3 2 1
1TR/1201/TIMS/*UD UNV/128MA

LIBRARY OF CONGRESS CATALOGING-IN-PUBLICATION DATA has been applied for.

ISBN 0-7534-5466-1

Printed in China

Series editor: Jackie Gaff
Series designer: David West Children's Books
Author: Amanda O'Neill
Consultant: Paul Hillyard
Editors: Claire Llewellyn, Clare Oliver
Art editor: Christina Fraser
Illustrations: Chris Forsey pp.4-5, 18-19, 30-31; Stephen Holmes 22-23, 26-27; Tony Kenyon (B.L.Kearley Ltd) all cartoons; Adrian Lascom (Garden Studio) 6-7, 10-11; Alan Male (Linden Artists) 14-15, 16-17, 28-29; Nicky Palin 20-21; Maurice Pledger (Bernard Thornton) 24-25; David Wright (Kathy Jakeman) 8-9, 12-13.

FOR CONNOR HOWARD O'NEILL

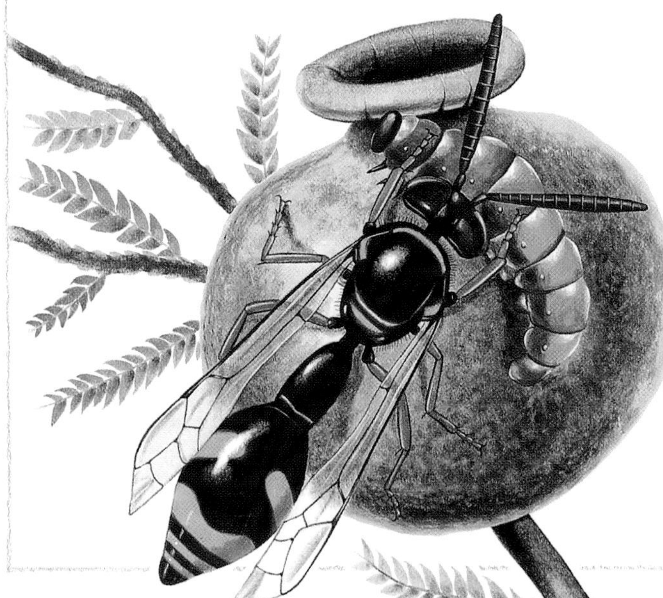

CONTENTS

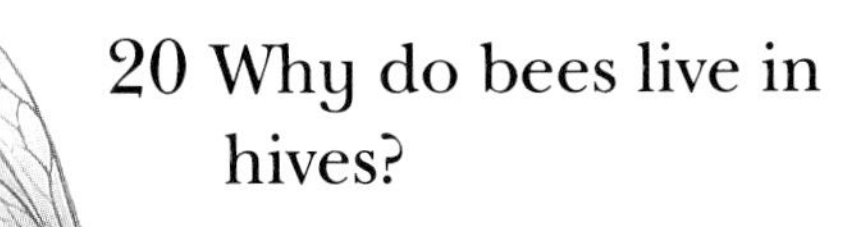

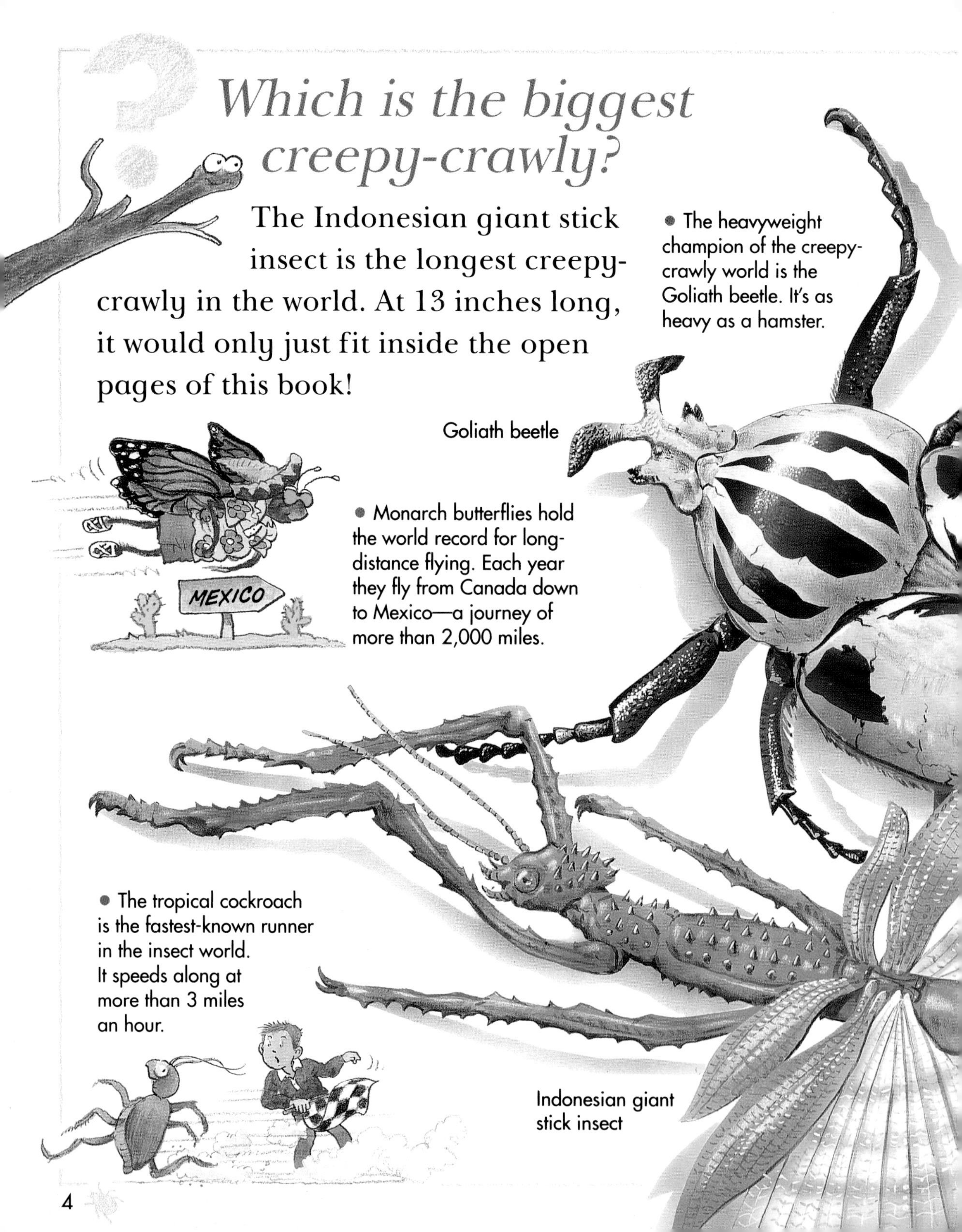

Which is the biggest creepy-crawly?

The Indonesian giant stick insect is the longest creepy-crawly in the world. At 13 inches long, it would only just fit inside the open pages of this book!

- The heavyweight champion of the creepy-crawly world is the Goliath beetle. It's as heavy as a hamster.

Goliath beetle

- Monarch butterflies hold the world record for long-distance flying. Each year they fly from Canada down to Mexico—a journey of more than 2,000 miles.

- The tropical cockroach is the fastest-known runner in the insect world. It speeds along at more than 3 miles an hour.

Indonesian giant stick insect

Which is the smallest creepy-crawly?

• Before the time of the dinosaurs, monster-sized dragonflies cruised through the air. Some were the size of seagulls!

You'd find it hard to see a fairy fly, because it's no bigger than a period. The female lays her tiny eggs inside the eggs of other insects. She can squeeze as many as 20 into just one butterfly egg.

• The Hercules emperor moth is the world's widest creepy-crawly. From wing to wing, it's the size of a dinner plate!

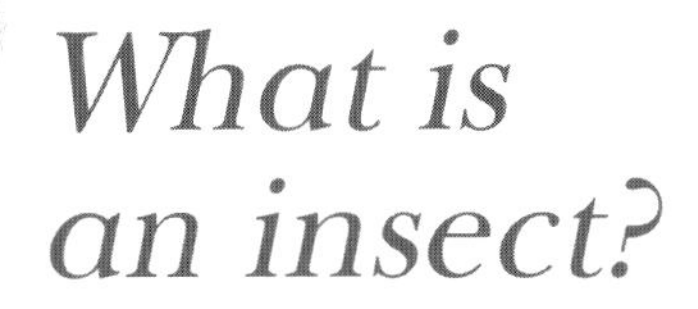

What is an insect?

An insect has three pairs of legs (that's six altogether), and three parts to its body. The first part is the head, the second is the thorax, and the third is the abdomen.

• Like all insects, this hoverfly has three pairs of legs and three parts to its body.

Feeler

Head

Eye

Thorax

Mouth

Leg

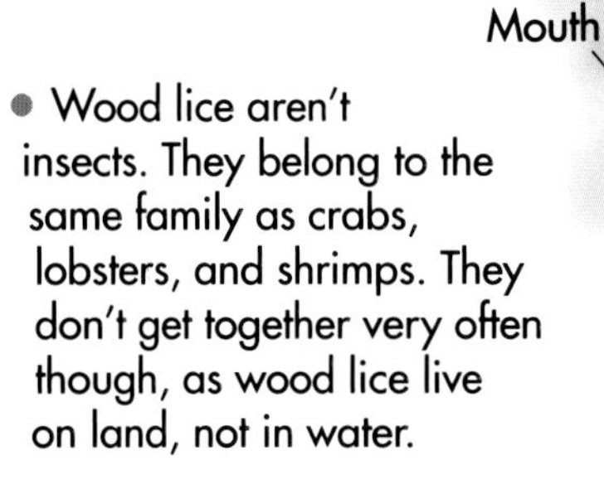

• Wood lice aren't insects. They belong to the same family as crabs, lobsters, and shrimps. They don't get together very often though, as wood lice live on land, not in water.

When is a fly not a fly?

A true fly, such as a housefly, has only one pair of wings. Butterflies, dragonflies, damselflies, and mayflies all have two pairs of wings. So they're not really flies at all!

What is a bug?

Bugs are insects that have needlelike beaks. A bug uses its beak to cut open its food. Then it sucks up the tasty juices inside, using its beak like a straw.

- Bedbugs are the draculas of the insect world. At night, they look for sleeping humans to bite. Then they suck up the tasty blood!

- Centipedes have too many legs to be insects. One kind has 176 pairs!

Are spiders insects?

No—a spider has eight legs, not six. What's more, its body has two parts instead of three. This is because the head and thorax are joined on a spider's body.

- There are more than a million kinds of insect—more than any other kind of animal in the world. And scientists are still finding new ones!

Why do spiders spin webs?

A spider's sticky web is its home and its pantry! When an insect flies into the web, it gets stuck. The spider rushes out to spin silk around it. Inside the silk, the insect turns into a liquid mush. Later the spider can suck it up like a drink!

• Most of the webs you see are round orb webs, spun by garden spiders. Other spiders spin webs with different patterns.

How do spiders make thread?

Spiders make runny silk inside their bodies, drawing it out through little knobs called spinnerets. The runny silk sets into thread, which the spider then weaves into a web, using its spinnerets like fingers.

• A Frenchman once took up spider farming. He soon gave the idea up, but not before he managed to make some spider-silk stockings and gloves.

• All spiders spin silk, but they don't all make webs. The spitting spider catches insects by spitting a sticky gum all over them.

When do spiders fly?

When baby spiders hatch out of their eggs, they spin themselves a long, silk line. Then they wait for the breeze to carry them off through the air to new homes, which may be many hundreds of miles away.

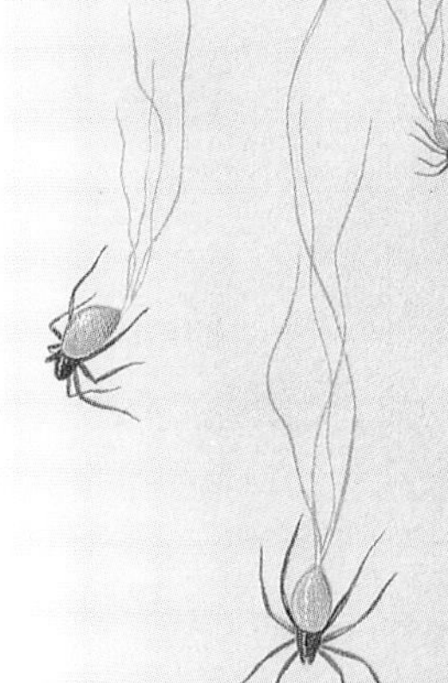

What does an insect feel with its feelers?

Insects use their feelers to taste and smell—the longer the feelers, the better. Some insects also use their feelers like fingers, to check things out by touching.

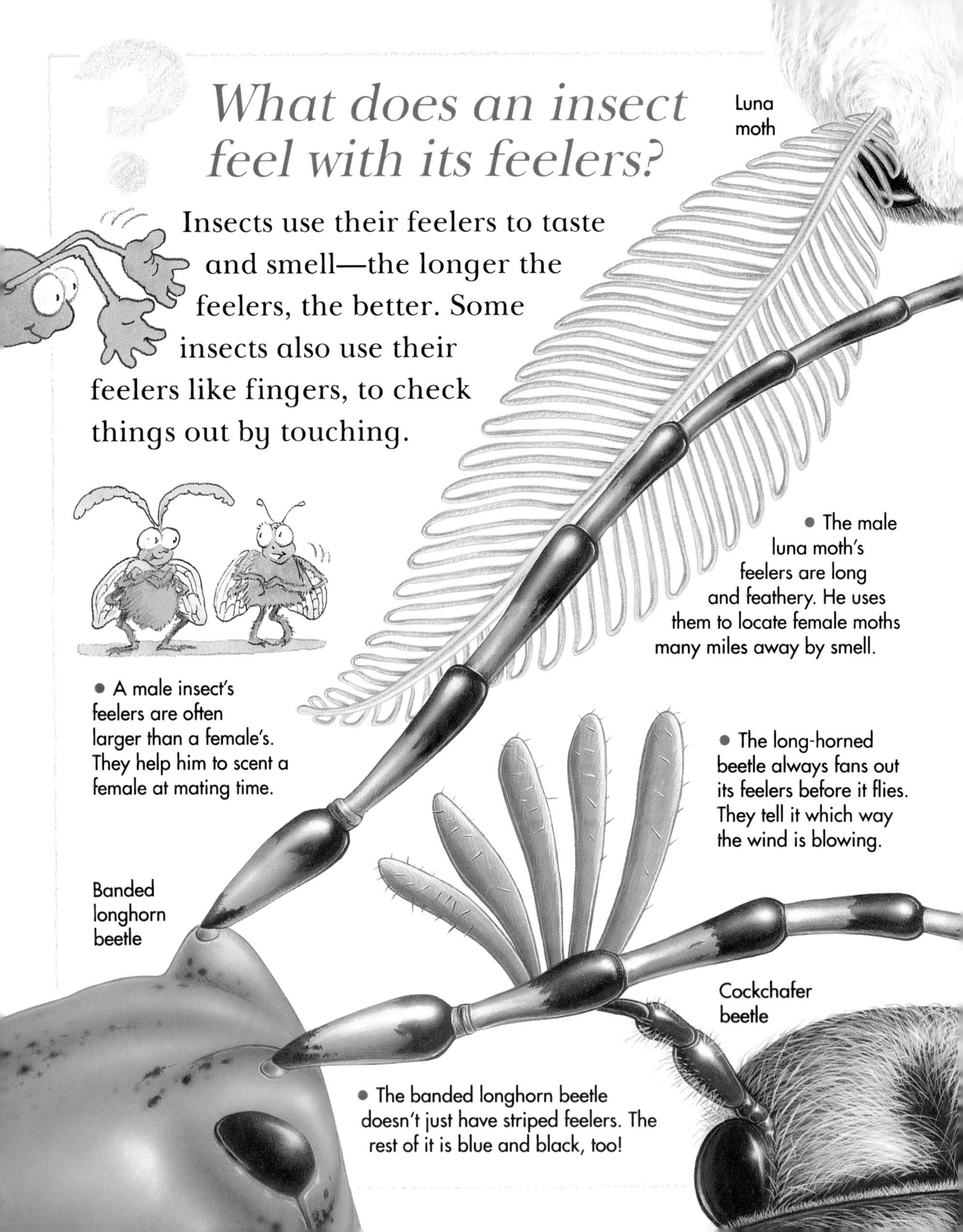

• The male luna moth's feelers are long and feathery. He uses them to locate female moths many miles away by smell.

• A male insect's feelers are often larger than a female's. They help him to scent a female at mating time.

• The long-horned beetle always fans out its feelers before it flies. They tell it which way the wind is blowing.

• The banded longhorn beetle doesn't just have striped feelers. The rest of it is blue and black, too!

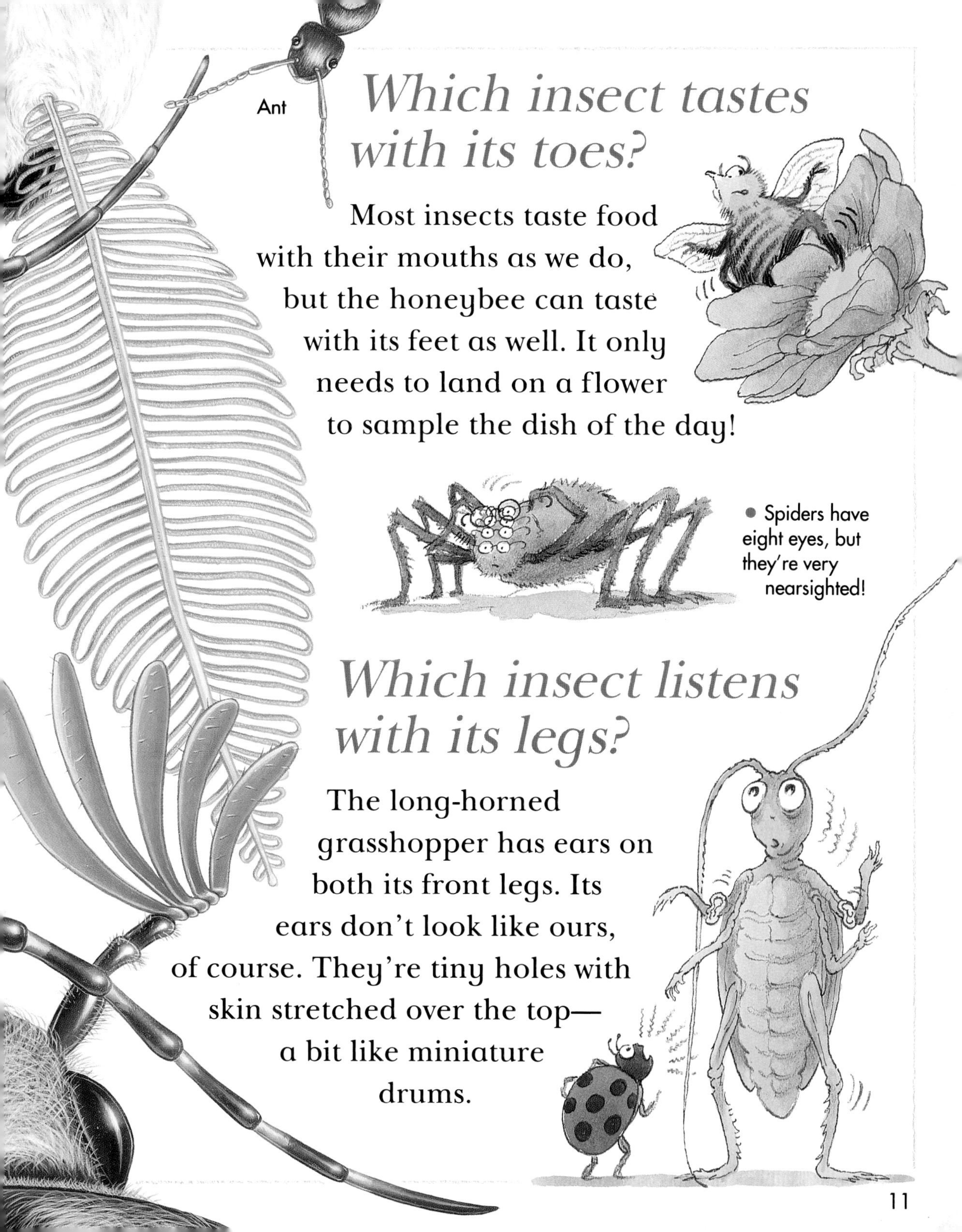

Which insect tastes with its toes?

Most insects taste food with their mouths as we do, but the honeybee can taste with its feet as well. It only needs to land on a flower to sample the dish of the day!

• Spiders have eight eyes, but they're very nearsighted!

Which insect listens with its legs?

The long-horned grasshopper has ears on both its front legs. Its ears don't look like ours, of course. They're tiny holes with skin stretched over the top—a bit like miniature drums.

Why do caterpillars change into butterflies?

Every butterfly has to go through four different stages of its life before it is a fully grown adult. At each stage, it changes its size, its shape, and its color.

• Many kinds of insect change shape as they grow. This way of growing is called metamorphosis.

1 A butterfly lays its eggs on a plant the baby caterpillars eat.

2 The caterpillars eat hungrily, and grow very quickly.

3 Each caterpillar makes itself a hard case called a pupa. Inside, its body turns into a kind of mushy soup.

• The babies that hatch from an insect's eggs are known as larvae—but many people just call them grubs.

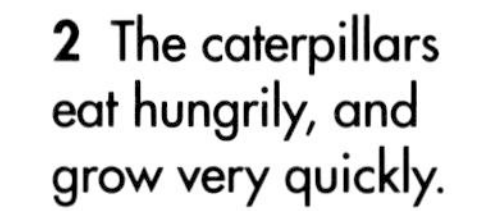

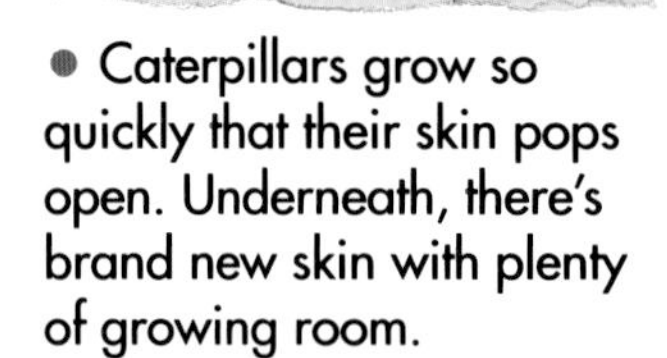

• Caterpillars grow so quickly that their skin pops open. Underneath, there's brand new skin with plenty of growing room.

• The pupa is like a strongbox. It keeps the insect's body safe while it changes shape.

• Nymphs split their skins as they grow, but they don't make a pupa. They just slowly change into adults.

• Not all insects change completely as they grow. A grasshopper's eggs hatch out into tiny nymphs, which look almost like their parents.

• A female butterfly lays as many as 50,000 eggs in her lifetime.

4 The soup slowly turns into a butterfly. When the butterfly wriggles out of the pupa, its wings are soft and creased. They dry in the sunlight.

• Butterflies don't need food to grow, but they love to sip sweet nectar from a flower now and then. It's a fuel that helps them to fly.

Which insect makes the best mother?

A mother earwig spends all winter looking after her eggs. She licks them clean and keeps them warm. When they hatch, she feeds them with food brought up from her stomach. Yummy!

• A wood louse carries her babies in a pouch, just as kangaroos do!

What hatches out in a vase?

The potter wasp makes tiny vases from wet clay. She lays an egg in each one and pushes in a live caterpillar, sealing the vase so that the caterpillar can't escape. When the baby wasp hatches out, its first meal is waiting for it!

Which insect is the laziest parent?

The cuckoo bee doesn't look after her own babies. Instead, she sneaks into another bee's nest and lays her eggs there. When the baby cuckoo bees hatch out, they eat any food they find—even their foster brothers and sisters!

• A queen bee lays up to 3,500 eggs a day, for several weeks at a time. No wonder she only lives for a couple of years!

• The larvae of the phantom midge float in the water on little pockets of air.

• The wolf spider carries her eggs around in a silk backpack. When the babies hatch out, they climb out onto her back and ride piggyback.

Which insects wear armor?

Beetles have two pairs of wings, but they only use one pair to fly. The other pair is like a thick piece of armor, covering the beetle's delicate wings and soft body.

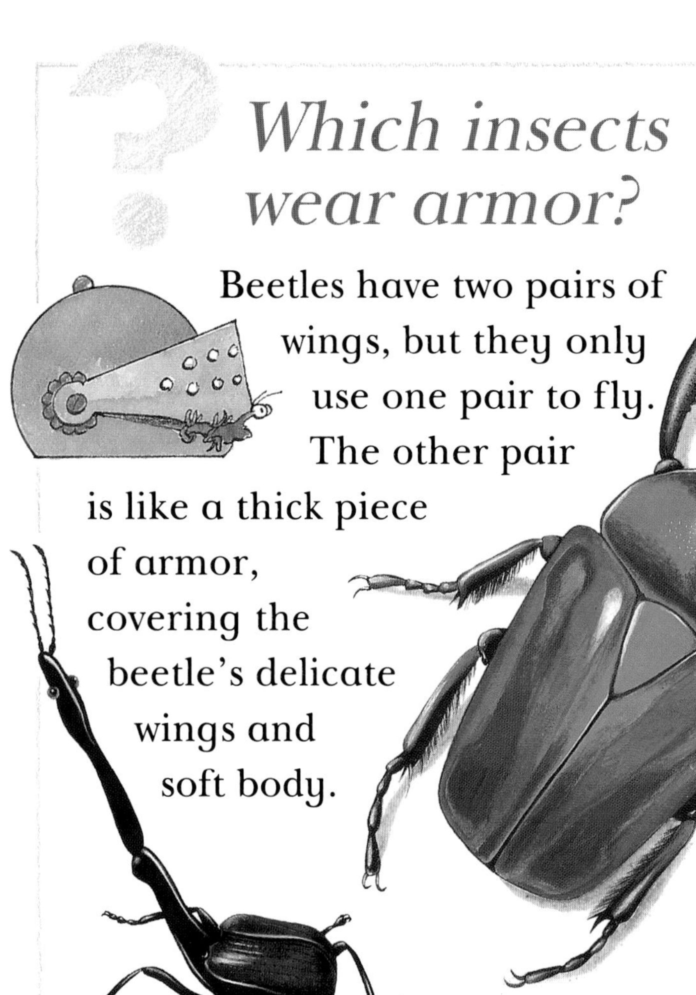

1 When a beetle is scuttling along the ground, its flying wings are hidden beneath a shiny armored wing casing.

2 When it wants to fly, a beetle opens its armor casing and stretches out its wings.

△ It's not difficult to guess how the giraffe weevil got its name. Its neck is twice as long as its body!

Which beetle fires a spray gun?

Watch out for the bombardier beetle! It shoots its enemies with a jet of hot, stinging liquid. As the jet is fired, it makes a sharp cracking sound like a tiny gun going off.

What digs graves, then robs them?

Burying beetles have no respect for the dead! When they find a dead animal, they dig away the soil until the body sinks into the ground. Then they lay their eggs inside the body and cover it with soil. When the eggs hatch out, there's a tasty, well-stocked meat store waiting for them!

• Imagine rolling balls of snow to make a snowman. That's how scarab beetles roll their balls of animal dung. They push the balls into a safe hiding place, and eat them later on.

• Ancient Egyptians thought the Sun was rolled across the sky by a giant scarab beetle.

3 The beetle beats its wings and moves smoothly through the air.

Why do moths flutter around lamps?

When moths fly at night, they use the Moon and stars to find their way. A bright lamp just confuses them. They circle and crash into it, and may even burn their wings on the hot bulb.

Why do glowworms glow?

A glowworm glows to attract a mate. It flashes out a message to other glowworms in the neighborhood. Then the glowworm sits back and waits for a reply!

• The female North American firefly tells lies. She flashes the signal of a different kind of firefly, and when the male comes near, she eats him!

How do grasshoppers play the violin?

The meadow grasshopper plays its body like a violin. It uses its back leg like a bow, scraping it against its wing to make a loud chirping sound. On a warm summer's day, you may have a whole orchestra in your yard!

• Deathwatch beetles like to munch through the floorboards, and they attract a mate by tapping their jaws. People once thought this knocking sound was a sign that someone in the house was about to die.

How do mosquitoes hum?

Female mosquitoes hum by beating their wings up to 1,000 times a second. They do this to attract a mate. The hum has a different effect on people, though. They rush to put on their insect repellent!

• The world's loudest insect is the cicada. You can hear it more than 500 yards away—that's about the length of five football fields!

Why do bees live in hives?

In the wild, families of honeybees live in holes in a tree or a rock. But if a friendly beekeeper provides a couple of cozy hives, then the bees are very happy to move in. This is really to help the beekeeper, of course—collecting honey from a hive is much easier than climbing a tree!

• When beekeepers open a hive, they wear special coveralls, gloves, and veils to protect themselves from beestings.

Why do bees dance?

When a honeybee finds plenty of food, it flies back to the hive to tell its friends. It does this by dancing. The number of wiggles in the dance and the direction the bee points tell the other bees exactly where to go.

Why do bees carry brushes and baskets?

Honeybees have little baskets on their back legs and brushes of hair on their other legs. When a bee lands on a flower, it brushes a powder called pollen into its baskets. Then it flies back to the hive to feed the pollen to its young.

- Bees not only have a brush, they have a comb! It's on their front legs, and they use it to clean their feelers.

- Collecting nectar to make honey is hard work. A bee would have to make 10 million trips to get enough for a single jar!

- Bears rob honey from bees' nests. Their thick fur protects them from bee stings —except on the nose!

- Not all bees like to live in hives. Many kinds of bee live alone, in burrows.

Which ants live in a tent?

Weaver ants sew leaves together to make tents for themselves, and they use their larvae as needles and thread! Each ant holds a young grub in its mouth, and pokes it through the edges of two leaves. The grub makes a sticky, silky thread, which stitches the two leaves firmly together.

Whose house has a trap door?

The trap-door spider's burrow has a door with a silk hinge which can open and shut. The spider hides inside, waiting for passing insects. When it hears one, it flings up the trap door and grabs its victim.

• Many creepy-crawlies make their homes in your home. Ants, spiders, moths, centipedes, and houseflies all like to live indoors.

Whose nest is paper thin?

The paper wasp's nest has paper walls. It makes the paper by chewing up strips of wood, which it tears from plants or old fence posts! It spreads the mixture in thin layers to build the nest.

• A tent caterpillar spins a shady silk canopy, and shelters under it while it feeds.

• Termites are champion builders, and make mud nests up to four times taller than a man. They need the room—as many as 5 million termites may be living inside!

Which dragon lives under water?

A baby dragonfly does! Young dragonflies are called nymphs. They can't fly, and they swim around in ponds and streams for a year or two while they grow.

3 Colorful adult dragonflies dart backward and forward above the water, their gauzy wings flashing in the sunlight.

2 When a nymph is ready to turn into an adult, it crawls out of the water onto a reed. Its skin splits open, and a dragonfly crawls out.

1 Dragonfly nymphs are fierce hunters, attacking anything that moves.

Which bug does the backstroke?

The greater water boatman spends all day lying on its back under the surface of the water, rowing along with its back legs. It must get tired of looking at the sky!

• The swamp spider loves fishing. It dabbles its feet in the water as bait for tiny fish. When they start to nibble its toes, it grabs them!

• The diving beetle's larvae are called water tigers, and they're fierce enough to kill a fish!

• The water spider makes an underwater tent of silk. To stop the tent floating away, the spider anchors it with silk threads.

When is a flower not a flower?

When it's a flower mantis! This insect is not only as pink as a flower, its body has knobbly parts that are shaped like petals. When an insect lands on the "flower" to feed, the mantis grabs it and eats it up.

• The treehopper is a tiny bug with a sharp disguise. When it sits on a twig, the shape of its back makes it look just like a thorn.

• If a bird swoops on a peacock butterfly, it gets a nasty shock! When the butterfly's wings open, they look just like a fierce pair of eyes.

When is a leaf not a leaf?

Leaf insects disguise themselves as leaves to hide from their enemies. Their green, leaf-shaped bodies are even crisscrossed with lines that look like the veins on a real leaf.

◁ This bee's black and yellow stripes mean, "Danger! I sting!" Hoverflies don't sting—but they wear a bee's coat to pretend that they do.

Which insect has a fierce bottom?

The Australian hawkmoth's caterpillar scares off its enemies by waving its bottom at them! The back end of its body looks like a fierce face with two bright circles for eyes. An attacking bird soon flies off in a hurry!

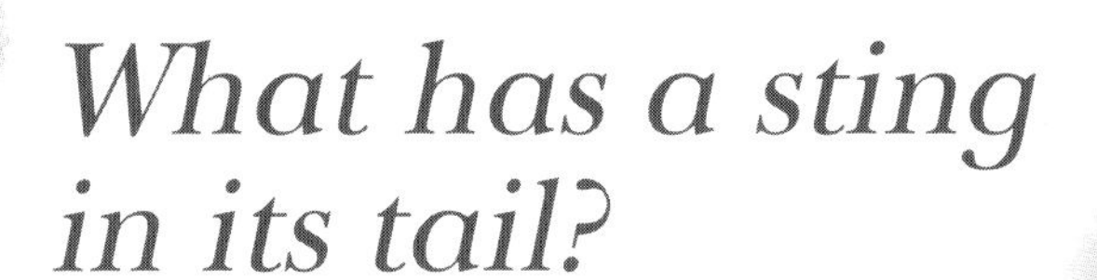

What has a sting in its tail?

A scorpion attacks by curving its tail over its head. It stabs its prey with the sharp end of its tail, and then squirts out poison. Ouch!

Can spiders kill people?

Although spider bites can hurt a lot, very few are poisonous enough to kill you. The black widow is the best-known deadly spider. Her poison is powerful enough to kill a human. She has another nasty habit, too—sometimes she eats her mate!

• The poison of a black widow spider is 15 times deadlier than a rattlesnake's.

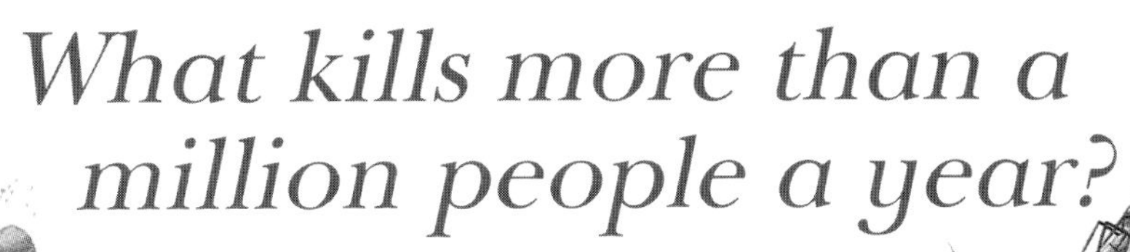

What kills more than a million people a year?

In some hot countries, mosquitoes carry a deadly illness called malaria. They can pass this on to anyone they bite. Between 1 and 2 million die from malaria every year.

• Doctors have cures for most spider bites, but they must know exactly which spider's poison they are treating. So if you're ever bitten by a spider, try to take it with you to the doctor!

Which killing machine has 120 million legs?

A swarm of army ants can be up to 20 million strong. The ants have no home, but live on the move, eating anything that gets in their way—and that can include people!

Are creepy-crawlies good for people?

Most creepy-crawlies are harmless —and some of them are valuable friends. They help plants to grow, keep the earth clean, and give us things such as silk to wear and honey to eat.

• People who study insects are called entomologists. They travel all around the world, learning about insects and trying to protect them.

• How would you feel about eating a toasted grasshopper? In different parts of the world, grasshoppers and caterpillars are served as tasty tidbits.

• Insects are the world's garbage collectors, eating up mucky things like dung and dead bodies.

• True locusts aren't good for people. They do more damage, more quickly, than any other insect. Locusts can gather in their millions and strip a crop field in minutes.

• Ladybugs are a gardener's best friend. They protect garden plants by eating huge numbers of hungry caterpillars and aphids.

• Rain forests are home to more insects than anywhere else in the world.

Are people good for creepy-crawlies?

Many creepy-crawlies gain by having us around. Our homes give them food, warmth, and shelter. But others have suffered because we've destroyed their homes or their food.

Index